FRANCYELLE SOUZA DE MENEZES RAMOS
RAIMUNDO RODRIGUES GOMES FILHO
ROSEANNE SANTOS DE CARVALHO

AGRONOMIC AND MICROBIOLOGICAL CHARACTERISTICS OF BUTTER CABBAGE

FRANCYELLE SOUZA DE MENEZES RAMOS
RAIMUNDO RODRIGUES GOMES FILHO
ROSEANNE SANTOS DE CARVALHO

AGRONOMIC AND MICROBIOLOGICAL CHARACTERISTICS OF BUTTER CABBAGE

IRRIGATED WITH TREATED WASTEWATER IN A PROTECTED ENVIRONMENT

ScienciaScripts

Cover image: www.ingimage.com

This book is a translation from the original published under ISBN 978-620-5-50439-0.

Publisher:
Sciencia Scripts
is a trademark of
Dodo Books Indian Ocean Ltd. and OmniScriptum S.R.L publishing group

120 High Road, East Finchley, London, N2 9ED, United Kingdom
Str. Armeneasca 28/1, office 1, Chisinau MD-2012, Republic of Moldova, Europe
Printed at: see last page
ISBN: 978-620-5-61343-6

SUMMARY

CHAPTER 01: INTRODUCTION

Water is a natural resource that is essential for the development of life on our planet. However, drinking water is becoming an increasingly scarce resource due to population growth. In recent years, irrigated agriculture has become one of the most important economic activities in Brazil. The large amount of water required for irrigation and the decrease in availability have increased interest in the rationalization of this resource, the lack of water resources, and the increase in conflicts over water use has raised concerns about conserving freshwater and intensifying the treatment of water already used for reuse.Wastewater reuse is an attractive option as it reduces the direct discharge of wastewater contaminants, thus improving the hygienic conditions of rivers, seas, aquifers and reservoirs. Areas of use for this type of water include irrigation of food or non-food crops, aquaculture, industry (water for cooling, cleaning) (MARTÍNEZ et al., 2012). The reuse of this effluent in agriculture, besides decreasing the use of good quality water, the wastewater can provide the nutrients required by crops, providing significant savings by reducing the use of chemical fertilizers, with a decrease in environmental impact due to the reduction of contamination of waterways, besides relieving the demand and preserving the water supply. Therefore the treatment of effluents is a necessity for the maintenance of the quality of water bodies.The benefits of reuse water from concentrated sewage in agriculture are numerous, reuse offers many advantages for the agricultural sector, despite the possibility of contamination risk, and the public health risks associated with pathogens that may be present in sewage water for reuse.With the increase in water demand for irrigation, the use of low quality water is an alternative to man and nature in the semi-arid region, reducing the demand for available water resources, as denoted by Lucena et al. (2018).Among the main factors limiting the survival and the economic and social development of a population throughout history is the lack of water. In a natural way it is possible to perceive the differences in the volume of water from region to region, linked to the lack of water. To this can be added the high interference of man in the environment, triggering an even greater hydrological imbalance (RAGAZZI, 2011).For agriculture, water quantity and quality are factors that directly influence crop production. In the search for sustainable agricultural production, water reuse becomes an alternative source of irrigation, because it reduces costs and expands the cultivated area and production, as reported by Lino et al. (2014).Once polluted water can be recovered and reused for beneficial purposes when its quality is maintained through adequate and safe treatment. Therefore, reuse has become a key word in terms of management in areas with low water availability or insufficient water resources, such as arid and semi-arid regions where water has become a limiting factor for the development of the industrial, urban and mainly agricultural sectors (HESPANHOL, 2003).An alternative of rationalization is the reuse of water for various uses, including irrigation, which not only saves but also has several other added benefits, including the recharge of the water table and the supply of nutrients, while respecting the sanitary and environmental limits of application to ensure the quality level of the vegetables (PAGANINI, 2003).Lima et al. (2012) report that the need for water varies between species during the cycle, thus knowing the behavior of species at each stage of development is of importance for correct management planning, taking into account the rational use of available water resources.Campos et al. (2015) state that water scarcity in arid and semi-arid regions has become increasingly recurrent, especially in developing countries and rural areas. Silva et al. (2014) state that the use of wastewater in agriculture shows itself as a valid alternative, since it makes available water necessary for the increase of agricultural production, besides reducing demand pressures on springs.Butter cabbage (Brassica oleracea L.) is an important vegetable grown in the West, being the species that most resembles the ancestral wild cabbage Brassica oleracea L. whose domestication is basically European (FILGUEIRA, 2008). It is an annual or biennial shrub vegetable in which the demand has been increasing considerably in Brazil, due to its great use in cooking and the new discoveries of its importance in nutritional properties. It has anticarcinogenic potential, because its leaves are rich in in glucosinolates, and has a high content of flavonoids, vitamins, and mineral nutrients. It is a culture that develops well in colder regions, typical of autumn-winter, presenting certain resistance to heat. Cabbage is consumed mainly in the form of salads, using the whole fresh leaves, and also in the composition of different dishes, sauces, soups, etc.The theme of the book therefore has a great relevance because it deals with the reuse of water inagriculture, as the largest volume of water is allocated to irrigation, the reuse of wastewater for agriculture is an alternative to optimize water resources and relieve the demand and preservation of water supply for other uses. In view of the above, the objective of this research was to analyze the agronomic and microbiological characteristics of butter cabbage subjected to different levels of irrigation with treated wastewater in a protected environment.

CHAPTER 02: THEORETICAL REFERENCE

WATER REUSE

The development generated by man has shown that the concept of water availability thought to be abundant and unlimited is not present today, which has led to the issues of water quality and economy becoming increasingly important in the context of environmental management and irrigation engineering. The increase in water and food consumption worldwide has been driven by the increase in consumption patterns and the growth of various human activities that, through various forms of pollution, affect the availability of drinking water, making it increasingly scarce, bringing water reuse as an alternative, being widely used in the agricultural area, which has stimulated several researches around the world (SANTOS et al., 2012).

The use of wastewater in agriculture brings advantages such as water reuse and reduced use of mineral fertilizers due to the application of organic matter and nutrients through irrigation systems and decreased use of good quality water. This will help to reduce potential pollution to the environment and disadvantages such as the occurrence of negative environmental impacts on the soil-plant system, contamination of soil, surface and groundwater and plant toxicity (RODRIGUES et al., 2011).

The use of municipal wastewater as a source of water for irrigated agriculture has received great interest (QADIR et al., 2010). The most important positive aspect of wastewater reuse in agriculture is that wastewater is available throughout the year, as it is not dependent on rainfall and seasons. This aspect allows the increase of irrigated areas, increasing the annual production of food and irrigation in places that are affected by the lack of water, mainly regions with arid and semi-arid climates (KERAITA et al., 2008).

As seen by Paganini (2003) the application of wastewater to the soil is a highly effective way to control pollution and a viable alternative for increasing water availability, especially in arid and semiarid regions, allowing the greatest benefits in economic, environmental and public health aspects. And that the reuse of water transforms a by-product of human activity, undesirable and even harmful, into a useful product, providing, in addition, other benefits such as increasing the nutritional value of plants. However, the use of treated wastewater in soils must be constantly monitored, to prevent contamination of the soil-water-plant system.According to Hespanhol (2003), the planned reuse of water is a potential alternative for the rationalization of this natural resource. The author highlights the importance of institutionalizing, regulating and promoting water reuse in Brazil, making sure that the practice is developed according to adequate technical principles, that it is economically feasible, environmentally sustainable, and socially accepted and safe, in terms of environmental preservation and protection of the risk groups involved.Irrigation with wastewater offers socioeconomic and environmental benefits, mainly the reduction of effluent discharge into water bodies and the recovery of nutrients (RODRÍGUEZ-LIÉBANA et al., 2014). However, depending on the origin and quality of the effluent, there are risks of soil and crop contamination by pathogens (PALESE et al., 2009), besides promoting the accumulation of salts, pH alteration and decrease in the infiltration rate of the soil (BEDBABIS et al., 2014).

Compared to surface water and/or groundwater the use of wastewater in irrigation increases biomass production and improves crop productivity (JANG et al., 2012; MOJID et al., 2012). Therefore, finding a balance between the use of these waters together with mineral fertilization is of great necessity, reducing the costs of fertilizers and providing a noble destination for these waters, increasing production, reducing spending on fertilizers and increasing the profitability of the producer.In the world, the reuse of sewage concentration plant effluents is growing every day, contributing to human and environmental sustainability, such as: improvement of the quality of life and socioeconomic conditions of rural populations, increase in agricultural productivity and recovery of degraded or unproductive areas, conservation and preservation of water resources, avoiding the discharge of raw sewage in springs (XAVIER, 2014).The application of sewage effluent to the soil is an effective form of pollution control and a viable alternative for increasing water availability in arid regions and semi-arid regions. The greatest benefits of this form of reuse are associated with economic, environmental and public health aspects (SILVA et al., 2011).

The use of treated wastewater in agriculture is important not only for serving as an extra source of water, but also of nutrients for crops (SANDRI et al., 2007). In this context, plants play an important role, which is to use the nutrients made available by wastewater, extracting macro and micronutrients, besides the carbon (organic matter) necessary for its growth, avoiding its accumulation and the consequent salinization of the soil and contamination of surface and groundwater (RIBEIRO et al., 2009).Whatever the form of reuse used, it is fundamental to observe that the basic principles that should guide this practice are: preservation of the users' health, preservation of the environment, consistent compliance with the quality requirements related to the intended use, and protection of the materials and equipment used in the reuse systems (HESPANHOL, 2002).Although in Brazil the practice of reuse is already being carried out in some states, mainly in São Paulo, there is no specific legislation dealing with the theme. However, as inducers of the beginning of the regulation process, working groups and technicians from the sector discuss and evaluate the issue in several meetings and national and international seminars, encouraging the institutionalization of recycling and reuse whenever possible, to promote the treatment and disposal of sewage, avoiding environmental pollution. Thus, there is also a need for regulation and the use of economic instruments for the control of water quality, with the purpose of increasing efficiency, thus reducing social costs and generating fiscal means to finance actions for the protection of the environment (BERNARDI, 2003).According to Dantas et al (2014), the effluent treated through stabilization ponds proved to be feasible in the irrigation of the radish crop showing no difference between treatments with respect to agronomic variables.Rebouças et al (2010) observed that increasing the proportion of maturation pond effluent in the supply water for fertigation of cowpea plants improves the dry matter production of root, stem and leaf, not only in one of these variables; evidencing the proportional growth of the plants. Sandri (2007) states that the use of wastewater proved to be a source of nutrients for elisa lettuce, interfering mainly in the formation of fresh mass and, consequently, in the leaf area.

WATER REPLACEMENT FOR CULTURE

Irrigated agriculture has been an important strategy to optimize the world's food production. In the past, irrigation was used only as a technique to apply water to combat drought. Today, it is one of the main focuses of agribusiness because it provides an increase in production, productivity and profitability of agricultural property (MANTOVANI et al., 2007). Faggion et al. (2009) state that irrigated agriculture has been using a much larger volume of water than necessary for satisfactory food production and that this reality is due to the inefficient use of water, such as the use of inadequate irrigation systems and the lack of determination of the necessary volume to be used for certain agricultural crops. As a decisive resource in food production, water needs to be increasingly used in a sustainable way, having an efficient use and alternatives that improve its use, such as the reuse of urban wastewater.

The shortage of water resources and increasing conflicts over water use have led to the emergence of conservation and treatment and reuse as formal components of water resources management. The inherent benefits of using reclaimed water for beneficial uses, as opposed to disposal or discharge, include preservation of high quality sources, environmental protection, and economic and social benefits (ASANO et al., 2007).

With the increase in the world's population, a greater production of food is necessary, increasing agricultural development and requiring new strategies for potentialization and lower risks in production. Thus, food production based only on the rainy season is not enough. One of the important challenges of today's agriculture is to increase competitiveness and product quality, associated with the preservation of water resources and the environment, allowing sustainable benefits on farms. Evaluate and adapt each of the factors that make up the production system, including the efficiency and management of irrigation water is an important factor, since agriculture is currently responsible for a large portion of the water used, making it necessary to implement efficient irrigation systems, in addition to using methods that quantify the water needs of crops, so that water and energy are not wasted (FACCIOLI, 1998).

According to WHO (2006), most of the water applied to the plant from rainfall and irrigation is consumed by the evapotranspiration process. Therefore, the water required by the crop is equivalent to the amount of water spent in this process, and the evapotranspiration rate depends on the type of crop and on climatic factors, which can be estimated according to meteorological data for the region in question.To determine crop water requirements, the most common method is based on the estimation of crop evapotranspiration (ETc), which is comprised of a two-step process. In the first, the evapotranspiration of a reference crop (ETo) is estimated. In the second, ETc is obtained by multiplying ETo by a crop coefficient (kc) that integrates the characteristics of the crop and the local climate (DOORENBOS et al., 1977).In consultation with various specialists, the FAO Penman-Monteith method was recommended as the standard method for defining and calculating reference evapotranspiration (ET0). This model was adopted because it presents relatively accurate and consistent results in both arid and humid climates (ALLEN et al., 1998). Doorenbos and Pruitt (1977), in general, divide the crop development cycle, for the calculation of crop coefficients, into four phases: initial phase with constant kc; crop development phase, in which the Kc increases linearly; mid-season phase with constant kc and the final phase, with a linear decrease. The kc assumes low values in the emergence phase and maximum values during the vegetative development period, which decline in the maturity phase.The realization of irrigation management depends on climatic variables obtained by methods that estimate reference evapotranspiration (ETo), a very important element for irrigation management, because it helps in the water consumption of crops. According to Gonçalves et al. (2009) ETo can be calculated by direct or indirect methods. O The direct method is the one that uses lysimeter, but even though it presents great results, the cost of the equipment is high, making its use impractical in daily irrigation management.

BUTTER CABBAGE CULTIVATION

Butter cabbage (Brassica oleracea L.) is an annual or biennial shrub vegetable in which the demand has been increasing considerably in Brazil due to its great use in cooking and the new discoveries of its importance in nutritional properties (NOVO et al., 2010).The cabbage stands out as one of the most important vegetables cultivated in the West, being the variety that most resembles the ancestral wild cabbage Brassica oleracea L. whose domestication is basically European. In this family the plants are attacked by several pests, such as aphids, cabbage curuquerê, cabbage moth, thread caterpillar and medfly caterpillar (FILGUEIRA, 2008).It belongs to the group of brassicas that do not form heads, and is mainly consumed in the form of salads, using the whole fresh leaves, and also in the composition of different dishes, sauces, soups, etc. (FELTRIM et al., 2003).It has anticarcinogenic potential, because its leaves are rich in glucosinolates, and it has a high content of flavonoids, vitamins and mineral nutrients (MORENO et al.,2006). It presents a large amount of carotenoids, with a high concentration of lutein and beta carotene, decreasing the chances of lung cancer and chronic eye diseases such as cataracts (LEFSRUD et al., 2007). It is a culture that develops well in colder regions (16 to 22° C), typical of autumn-winter, presenting some resistance to heat, depending on the location it can be planted all year round (FILGUEIRA, 2000). Cabbage is a culture that requires a good quantity of nutrients to be supplied by the soil. Among these nutrients, nitrogen is fundamental for its good development and maintenance of vigor.Harvesting is done every 7-10 days on the same plant, removing the well-developed leaves that are the size required by the market (20-30 cm long). These should be harvested by pulling their petioles (stalks) downward, with the The objective is to detach them near the point of insertion with the stem. After the harvest, the cabbage leaves are gathered in bundles with 8 to 12 units. This procedure can be done directly in the field or in a processing shed. The bundles are kept in water, to avoid withering, until marketing (TRANI, 2015).The average productivity of cabbage is 3 to 5 kg of leaves per plant, during a cycle of 6 to 8 months. The commercialization is done in the form of bundles of approximately 400 g or in the semi-processing system, where the leaves are chopped, sanitized and packed in trays, which adds greater value to the product (TRANI, 2015).Cabbage seedlings developed best in organic substrates, being worm humus and bovine manure (COSTA et al., 2011). Leafy cabbage has been propagated vegetatively by planting the lateral shoots that develop on the stem. The shoots are previously rooted in nurseries, with clay and very fertile soils, planting at a spacing of 15x15 cm (FILGUEIRA, 2008).

MICROBIOLOGICAL ASPECTS

According to Carvalho (2012) when wastewater is used for irrigation, the most important contaminants for public health are the biological ones. The determination of the amount of pathogens present in wastewater is of paramount importance due to the high risk that its use can bring to public health. It is important to emphasize that one should always promote an efficient treatment of the effluent to be used, choice and appropriate management of the irrigation system, restriction of the type of crop to be irrigated and care in harvesting, transport and handling. It also points out that the accumulated knowledge about the agricultural use of wastewater from sewage treatment plants in Brazil is in small steps, which makes it essential to research and actions towards controlled reuse, including its regulation, because the failure to adopt these criteria can lead to the indiscriminate use of treated wastewater for irrigation of various crops, and therefore a major vector for the spread of environmental pollution and disease.When using wastewater for irrigation, the first consideration is whether or not it contains pathogens, i.e., making sure that these pathogens are not pathogenic. microorganisms are not present in the water in densities that pose a significant health risk to users. The forms of control range from the application of effective concentration processes to the monitoring of water quality through analysis. Therefore, sanitary control is an essential factor of extreme relevance in the use of this technique, and care must be taken regarding contamination and the risks concerning public health, which is associated with pathogens that may be present in sewage water for reuse (WANDERLEY, 2005). Pathogenic and spoilage microorganisms can contaminate vegetables during pre- and post-harvest. During pre-harvest, the main sources of contamination are: soil, irrigation water, water used to apply fungicides and insecticides, dust, insects, inadequate composting, and human handling. The sources of post-harvest contamination include human handling, harvest equipment, transport packaging, animals, insects, dust, washing water, ice, transport vehicles and equipment during the process (BRACKETT, 1999; BEUCHAT, 2002).According to Wanderley (2005) the biological contaminants of importance to public health when treated wastewater is used for irrigation are pathogenic worms, protozoa, bacteria and viruses. Among the most relevant aspects of the use of wastewater for productive purposes, public health is still the subject of controversy in the international technical-scientific community. Allied to the cultural issue, these are factors that are active in the persistence of controversies in relation to the permissible risks and, consequently, in relation to the quality of the effluents, necessary and sufficient to guarantee health protection. As a way of evaluating the risks of the practice of irrigation with wastewater, it is necessary to consider the efficiency of the concentration processes used in the removal of pathogens through the use of indicator organisms, such as microbiological analysis of the water and the irrigated crop.The microbiological analyses are used to verify the organisms present and to quantify them, and can measure the risks that a certain foodstuff can offer to the consumer's health.

The analyses are of total importance to verify that the microbiological standards and specifications for food are being adequately met (FRANCO et al., 1996).For Hespanhol (2003), just the presence of pathogenic organisms in the wastewater, soil, or crops does not mean that disease transmission will occur. This is due to protective barriers, provided by factors characteristic of microorganisms (effective dose, persistence, residual burden, latency, etc.), of the hosts (natural or acquired immunity, age and sex, general health conditions), and other factors, which make the actual risk of causing disease generally much lower than the potential risk, characterized by the mere observation of the presence of pathogens.Salmonella is a genus of the family Enterobacteriaceae, Gram-negative bacteria, presented in the form of short rods and has a complex structure of lipopolysaccharides (LPS). It multiplies at temperatures between 7oC and 49.5oC, with 37oC being the optimum temperature for its development. They are appointed as the main etiological agents responsible for causing outbreaks in foods with high moisture content and high percentage of protein (YAMAGUCHI et al., 2013).Still according to Silva et al. (2010), Salmonella is a bacterium that occurs widely in animals and, in the environment, the main sources are water, soil, animal feces, insects, etc. The disease is usually contracted through the consumption of contaminated food of animal origin, especially beef, poultry, eggs and milk, and vegetables contaminated with manure can lead to transmission.

In Brazil, Resolution RDC No. 12 of the National Health Surveillance Agency of the Ministry of Health (ANVISA) of January 2, 2001, which regulates the sanitary microbiological standards for food, determines the maximum value for the presence of coliforms at 45°C and the absence of Salmonella sp in 25 grams of the sample.Dantas et al. (2014), verifying the viability of treated effluent through stabilization ponds in radish irrigation, stated that the values found for pathogens were within the standards accepted by ANVISA Resolution No. 12 of 2001.

CHAPTER 03: METHODOLOGY

CHARACTERIZATION OF THE EXPERIMENTAL AREA

The experiment was conducted in a protected environment, located in the Department of Agronomic Engineering (DEA) of the Federal University of Sergipe (UFS), in the municipality of São Cristóvão, Sergipe, under the geographic coordinates of 10º55'46 "S latitude and 37º06'13 "W longitude, at an altitude of 8 m. The experiment was conducted from September 29, 2019 to January 22, 2020 (Figure 1).

Figure 1 - Protected environment used in the experiment (DEA/UFS)

Source: Personal collection.

Shade cloth was installed inside the protected environment to mitigate the effects of sunlight and protect the plants from excessive heat and direct sunlight (Figures 2 and 3).

Figure 2 - Installing the shade cloth

Source: Personal collection.

Figure 3 - Sombrite already installed

Source: Personal collection.

An automatic weather station, model E5000 from the manufacturer IRRIPLUS (Figure 4), was installed inside the protected environment to monitor climatic data such as temperature, relative humidity, solar radiation and wind speed. These meteorological variables were used to estimate the crop water demand.

Figure 4 - Automatic Weather Station

Source: Personal collection.

The crop water demand was estimated according to Equation 1 and the reference evapotranspiration was estimated daily, using the standard FAO 56 Penman-Monteith method that was used initially (Equation 2). The crop coefficient used was the one recommended by FAO 56.

ETc = Kc x ETo(1)

0.408 x Δ(Rn-G)+Y 900 x U2 (es-ea)

ETo =T+273

Δ+Y x (1+0.34 x U2) (2)

where the variables are:

ETo = Reference evapotranspiration, mm day ;$^{-1}$

Δ = slope of the saturation vapor pressure curve, kPa °C^{-1} ; Rn = radiation balance at the surface, MJ m^{2} day ;$^{-1}$
G = ground heat flux, MJ m^{2} day^{-1} ; γ = psychometric constant, kPa °C ;$^{-1}$
T = air temperature measured at two meters height, °C;

U2 = wind speed measured at two meters height, m s^{-1} ; es = water vapor saturation pressure, kPa;
ea = actual water vapor pressure, kPa.
After a week of experimentation it was observed that the plants were presenting water deficiency with the Penman-Monteith method, because the blade applied was insufficient to meet the needs of the crop, observing symptoms of water deficit with completely wilted leaves and the soil was always dry. Then the irrigation method used changed to the Hargreaves-Samani method. This method is indicated for arid and semi-arid regions similar to the climate of the greenhouse, and is considered a simpler and more practical method for estimating reference evapotranspiration.

$$ETo = 0.0023(Tmed + 17.8)\ (Tmax - Tmin)^{0.5} \times Ra \qquad (3)$$

where:

T_{med} , $T_{max,}$ T_{min} - average, maximum and minimum temperatures respectively (°C);

R_a - solar radiation at the top of the atmosphere (mm day^{-1}).

The experimental design used was in Randomized Blocks (DBC) composed of fifteen treatments, three types of water T1 (100% water from the utility supply);

T2 (50% water from the utility supply + 50% wastewater); and T3 (100% wastewater), five sheets of irrigation being L1 (50% of ETc), L2 (75% of ETc), L3 (100% of ETc), L4 (125% of ETc) L5 (150% of ETc) in three blocks (three repetitions), totaling 45 pots (Table 1) (Figure 6).

Table 1- Experimental design

ETc	T1 (100% water)	T2 (50% water + 50% wastewater)	T3 (100% wastewater)
L1 (50%)	T1L1	T2L1	T3L1
L2 (75%)	T1L2	T2L2	T3L2
L3 (100%)	T1L3	T2L3	T3L3
L4 (125%)	T1L4	T2L4	T3L4
L5 (150%)	T1L5	T2L5	T3L5

Source: Own authorship, 2019.

Figure 6 - Overview of the experiment setup

Source: Personal collection.

WATER SOURCES

This experiment relied on two different sources of water: drinking water from the water utility, collected daily from a tap, located inside the protected environment, and treated wastewater from the ETE Rosa Elze, located in São Cristóvão-SE, collected and transported weekly to the experiment site in 5 liter plastic containers with lids (Figure 7).

Figure 7 - Collection of the treated wastewater from the Sewage Treatment Plant

Source: Personal collection.

The WWTP used in the study was fed by sanitary sewage in two points: one in the primary facultative lagoon, which represents the largest contribution of the system, according to DESO's information, receiving the sewage coming from the pumping station; the other in the secondary facultative lagoon, which receives the sewage by gravity. In both points, the sewage arrived at the pre-treatment unit, composed by grating and sand box, being then forwarded to the lagoons (Figure 8).

Figure 8 - ETE pre-concentration unit

Source: Personal collection.

EXPERIMENTAL CONDUCT

The soil was collected in an area of the Rural Campus of the Federal University of Sergipe. The collection was made in the 20 cm layer, as a way to also meet the demands of the culture. In the protected environment, the soil was sieved, homogenized and a sample of about 1000 grams was taken for chemical analysis by the certified soil laboratory of the Technological and Research Institute of the State of Sergipe (ITPS). The remaining soil was placed in plastic pots of 4 L each, which were properly positioned on metal benches in the center of the protected environment (Figure 9).

Figure 9 - Pots with soil

Source: Personal collection.

According to the results of the chemical analysis of the soil, liming was necessary; magnesian limestone was used, with a total neutralization power PRNT of 80%, which consisted in the application of 0.88 g vase^{-1} . As a way to improve the incorporation of limestone in the soil, it started to be irrigated daily. Sowing was done in trays, seeking the possibility of uniform seedlings and a good development. The seeds used were from Isla Sementes Couve de Manteiga da Georgia (Figures 10, 11, and 12).

Figure 10 - Butter Kale Seeds

Source: Personal collection.

Figure 11 - 10-day-old seedlings

Source: Personal collection.

Figure 12 - Seedlings at 20 days

Source: Personal collection.

Fertilization was performed 3 days before transplanting, allowing the nutrients to be incorporated into the soil. The fertilizer was calculated according to the results of soil analysis and the fertilizer recommendation was adapted from other states to Sergipe, established by Sobral (2007). The planting fertilization was composed of 0.27 grams of urea, 3.33 grams of simple phosphorus and 0.62 grams of potassium chloride per pot (Figure 13).

Figure 13 - Performing the planting fertilization

Source: Personal collection.

At 30 days after sowing in the trays, one plant per pot was transplanted, when the seedlings already had 4 definitive leaves (Figure 14, 15, and 16).

Figure 14 - Seedlings ready for transplanting

Source: Personal collection.

Figure 15 - Transplantation

Source: Personal collection.

Figure 16 - One seedling per pot

Source: Personal collection.

The irrigation was initially performed only with water from the utility company, in all pots, twice a day. The volume to be irrigated was defined weekly, observing if the culture presented symptoms of water deficiency until the seedlings adapted to the pot and the environment (Figure 17).

Figure 17 - Cabbage at 10 days after transplanting

Source: Personal collection.

Fifteen days after transplanting, when the seedlings were already adapted, the differentiation of concentrations was made with water from the utility and wastewater, and the irrigation sheets were determined. All irrigation was done manually, with the help of 100 mL beakers and millimeter cups directly in the pot, similar to a surface irrigation, directly on the ground without wetting the leaves.The tubes for the wastewater were separate, specified only for wastewater irrigation, also using gloves in this irrigation (Figures 18, 19, 20, and 21).

Figure 18 - Plants already adapted after transplanting

Source: Personal collection.

Figure 19 - Cabbage at 15 days after transplanting

Source: Personal collection.

Figure 20 - Millimeter probes and cups used for irrigation

Source: Personal collection.

Figure 21 - Irrigation with treated wastewater from the Sewage Treatment Plant

Source: Personal collection.

The cover fertilization was performed in two stages, the first with 20 days and the second with 40 days after transplanting as fertilization recommendation adapted from other states to Sergipe, established by Sobral (2007) being performed only for concentrations T1 (100% water). In the cover fertilization used only 0.27g of urea pot^{-1} (Figure 22).

Figure 22 - Covering Fertilization

Source: Personal collection.

The harvest was carried out 80 days after transplanting. The agronomic variables analyzed were number of leaves, leaf length, fresh mass and root length. The cabbage leaves were counted in each pot and the leaf length (Figure 23) and the root length (Figure 24) were determined using a ruler graduated in cm. The leaves were manually removed and stored in plastic bags, then the leaves were sent to the soil remediation laboratory in the Agronomy Engineering Department to be completed the agronomic analysis, for weighing the fresh mass of each pot we used a precision balance of 0.01 g (Figure 25).

Figure 23 - Determination of butter cabbage leaf length using a graduated scale

Source: Personal collection.

Figure 24 - Determination of butter cabbage root length

Source: Personal collection.

Figure 25 - Weighing the fresh mass of butter cabbage

Source: Personal collection.

Analysis of variance (anova) was performed for all parameters using the statistical software R (R Core Team, 2019). For qualitative analyses when significance was verified, the Tukey mean test was performed at 5% probability and for quantitative analyses when significance was verified, regression was performed. The cabbage leaves were sent to the ITPS (Technological and Research Institute of the State of Sergipe) for microbiological analysis of thermotolerant coliforms (coliforms at 45 °C) and Salmonella. During the experiment, the waters were monitored daily using a Lovibond multi-parameter

meter, model senso direct 150 series, to perform simultaneous monitoring of water quality indicator parameters such as pH, dissolved oxygen and electrical conductivity. The equipment incorporates intuitive software, robust, easy-to-view display and a case with electrodes, calibration solutions and accessories that collects data in real time (Figures 26 and 27).

Figure 26 - Multi-parameter meter

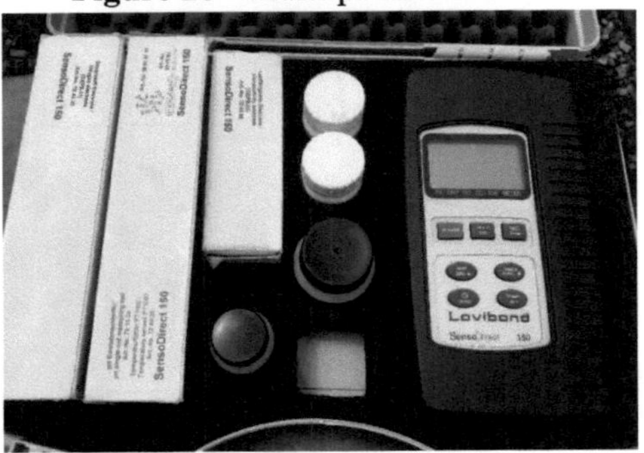

Source: Personal collection.

Figure 27 - Measuring pH, dissolved oxygen, and electrical conductivity of utility and wastewater.

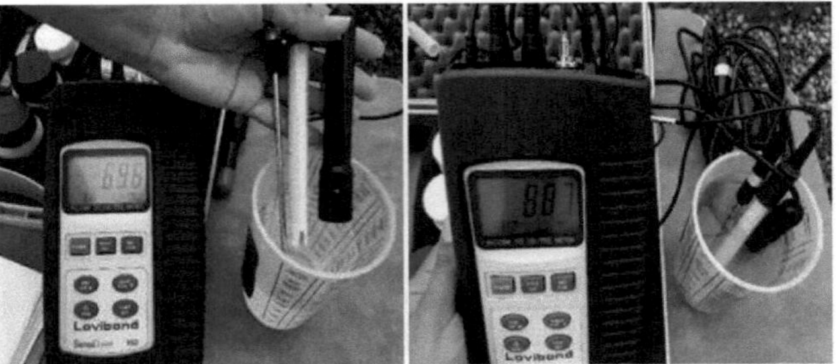

Source: Personal collection.

WEATHER CONDITIONS

Temperature

The experiment lasted 80 days, during this period the average air temperature was 29.78^0 C, and the maximum and minimum daily temperatures recorded were, respectively, 39.5 and 21.2^0 C. The average maximum and minimum daily temperatures during the cultivation period can be seen in Figure 28.

Figure 28 - Daily averages of maximum, minimum and average temperatures.

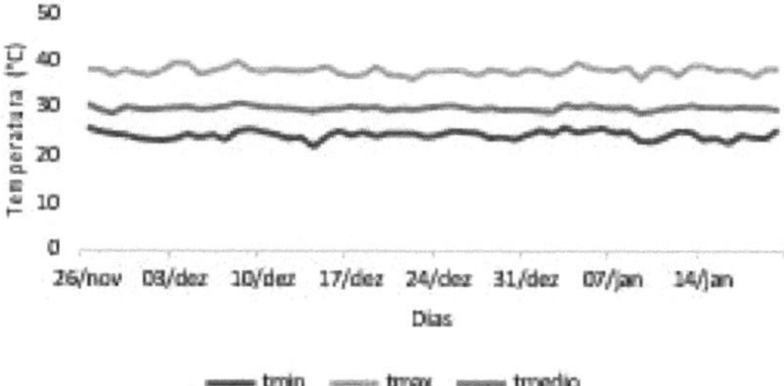

The relative humidity is a preponderant factor for reference evapotranspiration and consequently for the crop. During the experiment, the maximum and minimum relative humidity reached were, respectively, 78% and 68%, and the average daily humidity was 73%. Figure 29 shows the average daily relative humidity during the growing period.

Figure 29 - Daily averages of relative humidity.

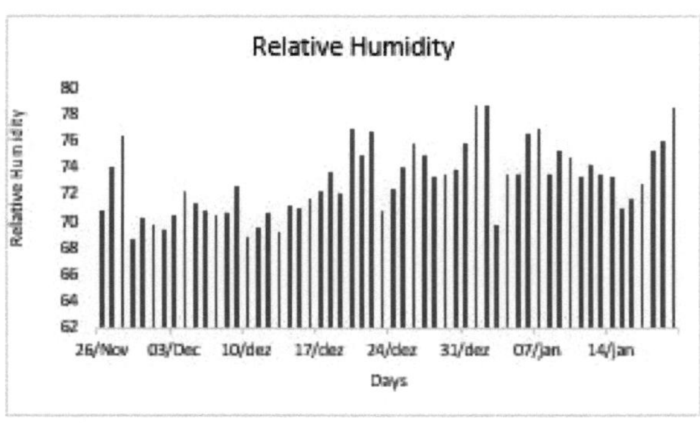

Solar Radiation

The solar radiation in the protected environment was monitored during the entire cultivation period. Radiation is a very important process for plant development. During the experiment, the maximum and minimum solar radiation achieved were 16.6 and 5.6 MJ m^{-2} , respectively. The average daily solar radiation was 9.18 MJ m^{-2} . Figure 30 shows the solar radiation during the experimental period.

Figure 30 - Daily averages of Solar Radiation.

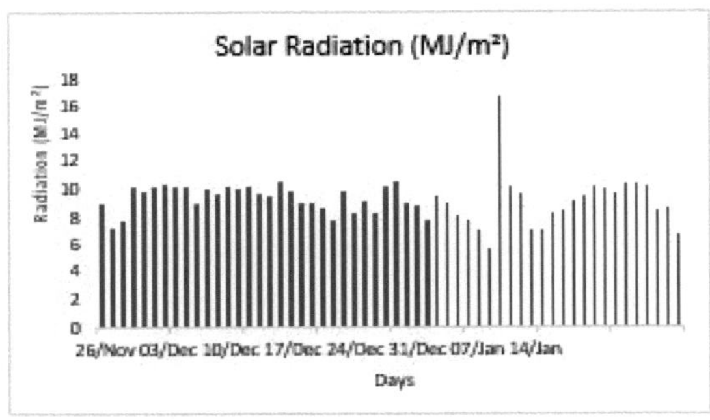

CHAPTER 04: RESULTS AND DISCUSSIONS

PLANT DEVELOPMENTS

After the first top dressing the plants developed well in the T1 concentrations and the plants that were irrigated with treated wastewater were also well developed. In all pots the cabbage leaves continued to grow uniformly. Photos of the development of the plants (Figures 31, 32 33, 34 and 35).

Figure 31 - Cabbage at 25 days after transplanting

Source: Personal collection

Figure 32 - Cabbage at 30 days after transplanting

Source: Personal collection.

Figure 33 - Cabbage at 35 days after transplanting

Source: Personal collection.

Figure 34 - Cabbage at 40 days after transplanting

Source: Personal collection.

Figure 35 - Cabbage at 45 days after transplanting

Source: Personal collection.

Sixty days after transplanting it was possible to notice the difference in the size of the cabbage leaves, due to the difference between the irrigation plates. It was noted that with the largest irrigation rates, the soil was always kept humid, favoring a better development of the leaves, and with the smallest rates the soil was not very humid, hindering the growth of the leaves (Figures 36, 37 and 38).

Figure 36 - Difference between the leaves

Source: Personal collection.

Figure 37 - Cabbage at 60 days after transplanting

Source: Personal collection.

Figure 38 - Cabbage at 70 days after transplanting

Source: Personal collection.

MICROBIOLOGICAL ANALYSES

The microbiological variables of the leaves were determined by the Technological Institute of Research of Sergipe (ITPS), for Thermotolerant Coliforms and Salmonella (the documents referring to the results issued by ITPS are in the appendices of this research). The absence of Salmonella and an amount lower than 3.0 NMP/g for thermotolerant coliforms were observed in the three concentrations applied (Table 2).

Table 2 - Results obtained for total coliforms (NMP/g) and Salmonella. Concentrations: T1(100% utility water), T2 (50% utility water + 50% wastewater),T3(100% wastewater).

Concentrations	T1	T2	T3
Coliforms (NMP/g)	≤ 3	≤ 3	≤ 3
Salmonella (absence in 25g)	Absent	Absent	Absent

Source: Own authorship, 2020.

According to the parameters analyzed for the Ministry of Agriculture and the Resolution - RDC No. 12 of 2001 ANVISA, the results obtained in the analysis of all samples met the established limits. Thus, it was possible to verify that the use of urban water reuse did not interfere with the microbiological characteristics of butter cabbage.The results obtained in the present study corroborated the observations made by Al-Nakshabandi et al. (1997) and Emongor (2004), when these authors found the absence of fecal coliforms, Salmonella sp., Shigela sp. and E. coli in all analyzed samples of eggplant and tomato irrigated with treated wastewater.According to Arman et al. (1994), lettuce, parsley, cabbage, onion, carrot and fennel plants presented high indices of contamination by indicator microorganisms after being irrigated with different types of sewage effluent through the sprinkler irrigation system. The count of these microorganisms varied according to the contamination of the water used, but was present in all the samples evaluated. This did not happen in the present study because the wastewater had no contact with the cabbage leaves.The results obtained in the present study also corroborated the observations made by Dantas et al. (2014) who irrigated radish with wastewater from the same WWTP and observed the absence of coliforms and Salmonella sp. Researches developed by Rego et al. (2005), with watermelon irrigation with treated effluent, showed, in all fruits tested, the absence of salmonellae and low values of fecal coliforms, independent of the irrigation systems used, attending, thus, the limits fixed by the National Agency of Sanitary Surveillance (ANVISA, 2001). Therefore, there was no compromise in the microbiological quality of the analyzed products, classifying them suitable for human consumption, similar to the results of this work with cabbage.According to the results of Varalho et al. (2011) the irrigation with wastewater showed efficient results in the lettuce culture, because the sanitary quality of the fertirrigated leaf was evaluated and the absence of thermotolerant coliforms was observed, results similar to those obtained in this work with cabbage.Souza et al. (2016) evaluated the microbiological quality of bell pepper fruits that were irrigated with wastewater and showed fruits free of microbiological contamination by coliforms and Salmonella, results consistent with those obtained in this work with the cabbage crop.Carvalho et al. (2013) who irrigated sunflower with wastewater, presented the results of analysis of thermotolerant coliforms and Salmonella in sunflower dry matter that were compared with legal standards of the National Agency for Sanitary Surveillance intended for human consumption and it was observed the regularity within the Brazilian Legislation, corroborating with the results in the present research with cabbage.Souza et al. (2015) verified that the reuse of wastewater in the culture of cowpea beans Brs Novaera did not present interference on the microbiological characteristics, as to coliforms and the absence of salmonella, verified under analysis having as parameter the guidelines of ANVISA.Souza et al. (2017) performed microbiological analysis of the okra crop irrigated with domestic wastewater and highlighting the results directed to the thermotolerant coliforms and salmonella found that are lower than the parameters established by ANVISA, so it is feasible the reuse of wastewater for irrigation of this crop, corroborating the results obtained in this study. Gomes Filho et al. (2020) observed that treated domestic wastewater can be considered a possible source of water for lettuce when applied to the soil in the vicinity of the crop root system, when they evaluated the microbiological quality of lettuce irrigated with different concentrations of treated domestic wastewater and irrigation depths. Silva et al. (2020), investigating the use of treated wastewater in the cultivation of beans in a protected environment, concluded that the application of water near the root system of the plants did not affect the microbiological quality of the beans. All treatments used were within the standards required by the National Health Surveillance Agency, corroborating the results obtained in this research.

AGRONOMIC ANALYSES

The agronomic variables analyzed were number of leaves, leaf length, fresh mass and root length. They were subjected to analysis of variance and analyzed by regression and Tukey's test at 5% probability level in R software (R Core Team, 2019).The analysis of variance showed that the factors blade and concentration separately were significant for the variables of leaf number (NF), leaf length (FC) The interaction between the factors blade and concentration was significant only in the variable of fresh mass (MF) and there was no significance of the factors studied for the variable of root length (CR) (Table 3).

Table 3 - Summary of the analysis of variance for leaf number (NF), root length (RC), leaf length (FC) and fresh mass (MF) of butter cabbage submitted to different irrigation slopes with domestic wastewater.

QM and significance					
FV	GL	NF	CR	CF	MF
Block	2	32,689	46,881	0,5336	51,1
Blade	4	37,144*	21,546ns	24,493*	3127,3*
Concentration	2	56,689*	1,924ns	17,5849*	3287,9*
Blade*Concentration	8	5,244ns	9,532ns	0,6618ns	457,7*
Waste	28	7,546	13,194	1,1526	82,9
Total	44	-	-	-	-
CV (%)		18,53	23,35	7,58	16,37

Note: GL: Degrees of freedom; ns: not significant; *: significant at 5% probability.

Source: Own authorship, 2021.

Dantas et al. (2014) found that the use of wastewater in the culture of beets showed no interference on the agronomic characteristics of the crop, where he also worked with wastewater from the same source used in this work (ETE of Rosa Elze) and found no significant differences at 5% probability by the Tukey test, results that differ from this work.Souza et al. (2017) performed agronomic analyses on the okra crop irrigated with wastewater in a protected environment and the results of the use of this water in the irrigation of the okra crop influenced the agronomic characteristics evaluated, corroborating with the results obtained in the present research.Santos et al. (2015) worked with the bean crop and when analyzing the agronomic characteristics of the aerial part, height, grain weight and number of pods, they observed that the non-performance of cover fertilization in the treatments with wastewater did not affect the development of the plant, therefore the nutritional contribution of wastewater to the plant was sufficient.Results of Magno et al. (2018) were dissimilar to the present research, where they studied the culture of maxixe in a protected environment and analyzed the agronomic characteristics irrigated with treated wastewater and reported that the use of this water did not affect the development of the maxixe fruit and did not show significant differences at 5% probability in any of the parameters analyzed.

Number of Leaves

According to the analysis of variance in Table 3, it was verified that there was a significant effect at 5% probability for the irrigation blade and for the concentration of wastewater alone and not significant for the interaction between them. In Table 4 the mean test of the concentrations for number of leaves can be observed.

Table 4 - Test of mean of concentrations of domestic wastewater for number of leaves of butter cabbage.

Concentration	Media	Result
T3	17,06667	A
T1	13,73333	B
T2	13,66667	B

Source: Own authorship, 2020.

According to Table 4, it could be observed that the concentration with 100% wastewater (T3) provided a greater number of cabbage leaves when compared to the treatments that did not apply wastewater (T1) and the one that applied 50% wastewater (T2).Similar results were found by Alves et al. (2012), who, when evaluating the possibility of using wastewater from domestic sewage in the production of tomato seedlings through the variables, number of leaves, leaf area, collar diameter, height of the aerial part and total dry mass, had as a result all the variables analyzed affected significantly by concentrations of wastewater, and the most vigorous seedlings were obtained in the highest concentration of wastewater in irrigation.In research developed by Freitas et al. (2012), there was a statistical difference for the number of leaves in relation to the type of water that was used for irrigation. According to the authors, the plants that were irrigated with sewage water presented an average number of leaves of 25.31 and 22.31, for those irrigated with water supply, corroborating the results obtained in the present research.Medeiros et al. (2020), worked with sunflower plants and observed that the values were not different, irrigated with wastewater and artesian well water showed an average value of 24.00 and 21.23 leaves, respectively. The treated domestic wastewater was the source of water for irrigation of sunflower plants that showed better results for all variables evaluated, equal to the results in this study.Silva et al. (2015), who worked with lettuce type crespa, found no difference ($p < 0.05$) in the number of leaves when using 100% treated wastewater. In a study of melon, Villela et al. (2003) found that the treatments derived from the use of residual sewage did not interfere in the number of leaves of the plant of this crop, both studies contradicted the results found in this study.Azevedo et al. (2005) verified that treated wastewater contributed effectively to a greater production of cotton in relation to the supply water, and the value was 65.98% higher in relation to treatments that received only irrigation water, corroborating the tendency to increase the number of cabbage leaves observed in this study. The polynomial model that best fitted the behavior of the number of leaves as a function of the irrigation sheets corresponding to the crop evapotranspiration was quadratic. Figure 39 shows a single regression curve of the average number of leaves. The largest number of leaves of cabbage was found using 122% of crop evapotranspiration, reaching 16 leaves.

Figure 39 - Average number of leaves of cabbage as a function of irrigation rates based on crop evapotranspiration.

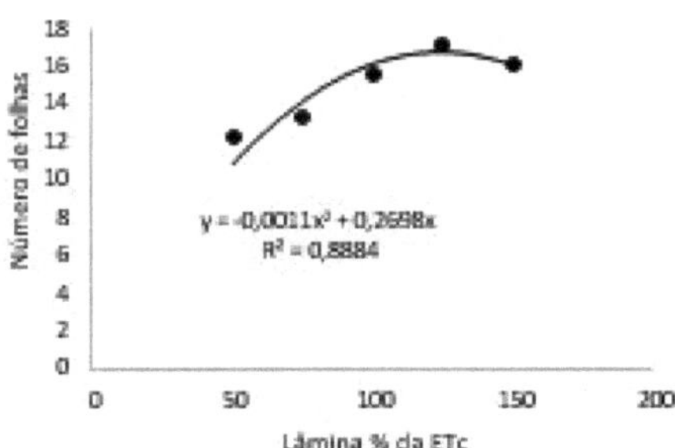

Santana et al. (2019) obtained greater amounts of lettuce leaves when blades corresponding to 125% of the crop evapotranspiration were applied, for a maximum total of 13 leaves. Cassimiro et al., (2019), on the other hand, recommended irrigation corresponding to 100% of the crop evapotranspiration for crisp lettuce for the climatic situations of Souza in Paraíba.Similar results were found by Silva et al. (2019) working with feijao caupi, within the protected environment observed that the result found demonstrated that plant height and other variables increase as the replacement blade is increased.

Root Length

According to the analysis of variance in Table 3, it was found that there was no significant effect at 5% probability for irrigation blade, wastewater concentration and for interaction among them for root length.

Leaf Length

According to the analysis of variance in Table 3, there was a significant effect at 5% probability for leaf length, in relation to the treatments irrigation blade and concentration of wastewater. According to Table 5, it could be observed that the treatment with 100% utility water (T1) and 100% wastewater (T3) provided greater length of cabbage leaves when compared to the treatment that applied 50% wastewater (T2).

Table 5 - Test of the mean of the concentrations.

Concentration	Media	Result
T1	15,16	A
T3	14,33	A
T2	13,01	B

Source: Own authorship, 2020.

According to Freitas et al. (2012) the vegetative performance of plant height and leaf length irrigated with reuse water is related to the macro and micro nutrients dissolved in the water. Souza et al. (2016) used treated wastewater in the irrigation of the bell pepper crop and influenced the agronomic characteristics evaluated. The use of treated wastewater in irrigation caused changes in agronomic variables performed with a significance level of 5%, similar results were found in this research.Carvalho et al. (2019) observed that for bell pepper production, the factors average fruit mass, fruit length, fruit diameter, and length-diameter ratio, irrigating with different water qualities and applying two irrigation managements, provided similar agronomic performance, which reinforces the technical feasibility of water reuse. Different results were found by Oliveira et al. (2014), who, when evaluating the use of wastewater in irrigation of sunflower crops, observed that it did not influence the quantity and length of leaves, and also evaluated the productivity of the crop through the amount of dry matter of the seeds and dry matter of the flower, stem, and leaf. Santana et al. (2019) analyzed the influences of the use of treated wastewater on the agronomic characteristics of cilantro in a protected environment and concluded that the use of wastewater did not show differences in the development of the leaves of cilantro. In this work the results showed a difference in the development of cabbage. The polynomial model that best fitted the behavior of leaf length as a function of the irrigation blades corresponding to the crop evapotranspiration was quadratic. Figure 40 shows the regression curve of average leaf length because the interaction concentration*blade was not significant analyzing all the repetitions of each blade. The greatest leaf length of cabbage was found using 120% of the crop evapotranspiration, reaching 15 cm.

Figure 40 - Average leaf length as a function of irrigation rate based on crop evapotranspiration

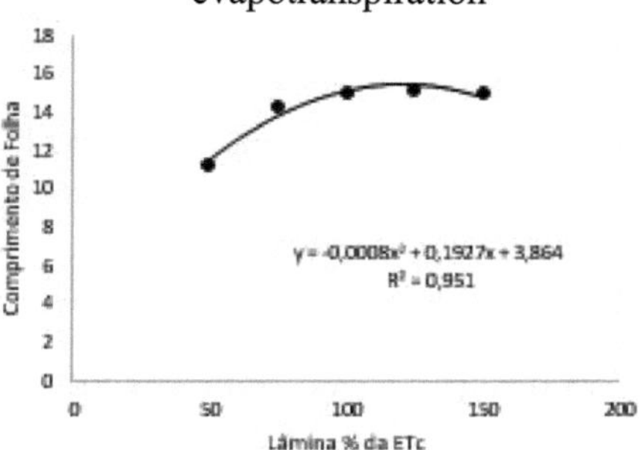

Fresh Pasta

According to the significant interaction between the factors for this variable, the figures below show the behavior of the type of water used within the blades used.According to the analysis of variance in Table 3, there were significant effects for blade and concentration separately and for the interaction between them. In Table 6 the mean test was performed for the interaction blade of irrigation x concentration of wastewater.

Table 6 - Summary of the slides within each concentration for fresh mass of cabbage

Blades	Concentration of wastewater		
	T1	T2	T3
L1	32,120 Ac	20,896 Ab	29,186 Ac
L2	49,233 Abc	42.916 Aa	49,906 Ac
L3	63,570 Ab	43,896 Ba	75,036 Ab
L4	88.456 Aa	44,223 Ba	88,336 Aab
L5	62,803 Bb	44,646 Ba	98,776 Aa

Note: Means followed by equal lower case letters in the columns and equal upper case letters in the row, did not differ by the Tukey test at 5% probability.
Source: Own authorship, 2021.

Observing the results of the influence of the blades in the concentrations, the blade L5 with the concentration T3 showed the highest fresh mass of cabbage. In the T1 concentration the greatest fresh mass was in the blade L4 where it presented difference among the other blades, the blades L2, L3 and L5 did not present differences among themselves, and the blades L1 and L2 also did not present differences. For T2 concentration, lamina L1 was the only one that showed differences from the other laminas, L2, L3, L4 and L5 showed no differences among themselves. For T3 concentration, blades L1 and L2 did not show differences among themselves, as well as L3 and L4. L4 and L5 did not show differences between them either.According to Dantas et al. (2014), also working with wastewater from the same source of the Rosa Elze WWTP found no significant differences at 5% probability by Tukey's test in radish culture, which differs from this work.Carvalho et al. (2019) worked with the production of bell bell pepper in a protected environment with wastewater and concluded that the productivity and physical characteristics of bell bell pepper fruits were not affected by the quality of irrigation water, results dissimilar to those found where the productivity of cabbage was higher when irrigated with wastewater.Souza et al. (2013) worked with beet and carrot crops in a protected environment and emphasize the importance of using domestic wastewater to supply nutrients and increase the productivity of the plants. It was found that the use of treated wastewater did not influence the carrot crop, and there was some influence on the beet crop.

Similar results were found by Andrade et al. (2012) who, worked with sunflower and two water qualities (supply and domestic wastewater),found that plants irrigated with wastewater showed 16.54% more growth than plants irrigated with the water supply.According to Cunha et al (2015), working with the cultivation of beans irrigated with wastewater there was no significant difference at the 5% level of probability by the Tukey test, so it did not influence the productivity of beans, different results of this work where it showed that wastewater caused greater fresh mass.In Figure 41, the model chosen for the trend line was the second order polynomial, which best fitted the behavior of the fresh mass of cabbage as a function of the irrigation sheets corresponding to the evapotranspiration of the crop. As the interaction concentration*blade was significant a graph representing the three concentrations was made.

The regression curves show the averages of each blade for each concentration. At concentration T1 (100% water supply) the maximum peak of fresh mass was obtained by applying 123% of ETc, with 74 grams.

The inflection point for concentration T2 was obtained by applying 112% of ETc, with a fresh mass of 47 grams. For concentration T3 (100% wastewater) the inflection point was obtained by applying 196% of ETc, with a fresh mass of 107 grams.

Figure 41 - Regression of fresh mass in the concentrations as a function of irrigation rates based on crop evapotranspiration.

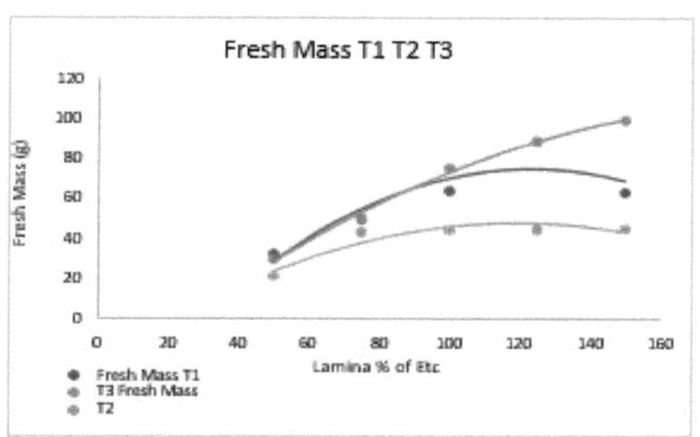

Nobre et al. (2010), using different percentages of sunflower crop water requirement, observed that the number of leaves, plant height and fresh phytomass and drought, increased as the percentage of water source supplied to the crop increased, corroborating the results in the present research with cabbage.Massaranduba (2020) stated in his work that the water stress caused by the deficit was noticeable for replacement slabs lower than 100% of ETc, progressively decreasing the commercial productivity of onion bulbs. The highest total and commercial yields were found at 122% of ETc. As in this research, a higher yield was obtained for T2 concentration using 112% of ETc. Similar results were found by Baptestini et al. (2018) growing onions of the hybrid cultivar Aquarius, obtained maximum total productivity of 60.7 t ha^{-1} , using blade of 150% of ETc, in the municipality of Viçosa-MG. It was observed in the present study that the increase in fresh mass in concentration T3 (100% wastewater) was greater when compared with the fresh mass in concentration T1 (100% water supply), so the cabbage obtained better and greater results when irrigated with treated wastewater.Ferreira et al. (2005), studying herbaceous cotton, verified that wastewater promoted greater plant growth in height and leaf area, in relation to the supply water.Noreto et al. (2012) noticed an increase in the diameter, fresh and dry mass of lettuce as the percentage of water applied increased, as well as observed in the present work when a better average for fresh mass was noted at concentration T3 (100% wastewater).

QUALITY OF THE WATER USED

The chemical analysis of the waters used in the research were comprised of: pH, dissolved oxygen (DO) and electrical conductivity (EC). They were compared by the limits of physical-chemical parameters of Class 1 of CONAMA Resolution 357/2005, which among the predominant uses includes the irrigation of vegetables that are consumed raw and fruits that develop close to the ground and are eaten raw. without removal of film (BRASIL, 2005a). And by Resolution CONAMA 430/2011, which complements and amends Resolution CONAMA 357/2005 and provides on the conditions and standards for effluent discharge.

pH

According to CONAMA Resolution No. 357/2005 for the effect of pH in water for irrigation, it is recommended values between 6 and 9. Therefore, by the data presented in Table 7, the water from the concessionaire and the wastewater were within acceptable standards for irrigation. The wastewater that has a high concentration of H+ ion, that is, low pH, is difficult to be treated biologically.

Table 7 - Average pH values

CONCENTRATION	PH
T1	7,6
T2	9,2
T3	9,2
CONAMA 357/2005	6,0 - 9,0

Source: Own authorship, 2020.

It should be emphasized that the importance of pH is not only in the biological and chemical reactions existing in sewage treatment, but when water is used for irrigation, the fact that pH is too acidic or too basic can lead to serious problems of nutrition and toxicity to the crop, as well as the appearance of incrustations and even corrosion in irrigation systems (DUARTE, 2006). Duarte (2006) pointed out that in relation to the effect of pH on water for irrigation, values between 6.5 and 8.4 are recommended. The concentrations of H^+ and OH^- obtained in irrigation water can influence the availability and absorption of nutrients by plants, the structure and properties of the soil and irrigation systems. Therefore, according to the values presented in the table, they are outside the range considered ideal for the author, but did not represent negative effects regarding the practice of irrigation.

Dissolved Oxygen

In this research, the analysis results for dissolved oxygen were outside the standard, as required by CONAMA nº 357/2005, according to the values that can be seen in Table 8.

Table 8 - Mean values of Dissolved Oxygen

CONCENTRATION	DISSOLVED OXYGEN
T1	4.6 mg L-1
T2	0.8 mg L-1
T3	1.1 mg L-1
CONAMA 357/2005	Not less than 6 mg L-1

Source: Own authorship, 2020.

According to CONAMA nº 357/2005 for class 1 fresh waters the value is not less than 6 mg L^{-1} for class 2 values not less than 5 mg L^{-1} for class 3 values not less than 4 mg L^{-1} and class 4 values higher than 2 mg L^{-1} . Therefore, for concentration at T1 the water was considered a class 3 water.According to Duarte et al. (2008), in irrigating with sanitary sewage treated by various treatment technologies, they concluded that the sewage used showed adequate physical and chemical quality for bell pepper plants.

Electrical conductivity

According to Mancuso (2003), the capacity of water to conduct an electric current is greater the higher the concentration of electrolytes, that is, the salinity of the reuse water can be measured by the electric conductivity. Ribeiro et al (2004) also confirmed that the salinity level can be measured by electrical conductivity, or even the concentration of soluble salts present in irrigation water.Electrical conductivity (EC) is a measure of the total concentration of dissolved salts present in water. The greater the amount of dissolved ions, the greater the electrical conductivity in the water. For Feitosa and Manoel Filho (2000), the electrical conductivity tends to increase due to several factors, among them, temperature rise and higher concentration of dissolved ions. According to Esteves (1998), the ions directly responsible for the electrical conductivity values are the so-called macronutrients, such as calcium (Ca), magnesium (Mg), sodium (Na) and potassium (K).In this research the analysis results presented values within the standard, as required by CONAMA No. 357/2005, according to the values that can be seen in Table 9.

Table 9 - Average values of Electrical Conductivity

CONCENTRATION	COND. ELECTRICS
T1	0.29 mS
T2	0.53 mS
T3	0.75 mS
CONAMA 357/2005	> 5.5 high salinity

Source: Own authorship, 2020.

CHAPTER 05: CONCLUSION

The results of the analysis of coliforms at 45 °C and Salmonella were lower than the parameters established by ANVISA (National Health Surveillance Agency). The use of treated wastewater for irrigation of butter cabbage influenced the agronomic characteristics evaluated, improving the development of the plant. With the use of treated household wastewater for irrigation of butter cabbage, a greater fresh mass of cabbage was obtained compared to the supply water, obtaining 107 grams of fresh mass when it was applied196% of ETc. The results of electrical conductivity showed values within the standard. As for pH, only the T1 concentration had a value within the standards, while the T2 and T3 concentrations were above the standards, passing 9.0. For Dissolved Oxygen, all concentrations presented values outside the standards.

REFERENCES

ALLEN, R. G.; PEREIRA, L. S.; RAES, D.; SMITH. M. Crop **evapotranspiration - Guidelines for computing crop water requirements**. In: FAO Irrigation and DrainagePaper 56. Rome: FAO, 1998.

AL-NAKSHABANDI, G. A.; SAQQAR, M. M.; SHATANAWI. M. R.; FAYYAD, M.;ALHORANI, H. Some envirommental problems associated with the use treated wastewater for irrigation in Jordan. **Agricultural Water Management**, 1997.

ANDRADE, L. O.; GHEYI, H. H.; NOBRE, R. G.; DIAS, N. da S.; NASCIMENTO, E. C.S. Flower quality of ornamental sunflowers irrigated with wastewater and supply water. **Idesia**, 2012.

ANVISA - Agência Nacional de Vigilância Sanitária (2001**) Resolution RDC nº 12, of January 2nd, 2001**. Approved the Technical Regulation on microbiological standards for food. Diário Oficial da União; Poder Executivo, January 10, 2001.

ARMAN, R. et al. **Residual contamination of crops irrigated with effluent of different qualities: a field study.** Water Science and Technology, 1994.

ASANO, T. **Water reuse, issues, technologies, and applications.** New York: Metcalf and Eddy/AECOM; McGraw Hill, 2007.

AZEVEDO, M. R. Q. A.; BELTRÃO, N. E. de M.; KONIG, A.; AZEVEDO, C. A. V. de;PORDEUS, R. V.; TAVARES, T. de L. Análise comparativa da produção do algodoeiro herbáceo irrigado com água residuária e água de abastecimento e adubação nitrogenada. **Congresso Brasileiro de Algodão**, 2005.

BAPTESTINI, J. C. M.; OLIVEIRA, R. A.; VIDIGAL, S. M.; PUIATTI, M.; CECON, P. R. Onion productivity in relation to irrigation water depths and nitrogen doses. **Horticultura Brasileira**, 2018.

BEDBABIS, S.; ROUINA, B. B.; BOUKHRIS, M.; FERRARA, G. Effect of irrigation with treated wastewater on soil chemical properties and infiltration rate. **Journal of Environmental Management**, London, 2014.

BERNARDI, C. C. **Reuse of water for irrigation**. Brasília: isaefgv /ecobusiness school, 2003.

BEUCHAT, L. R. **Ecological factors influencing survival and growth of human pathogens on raw fruits and vegetables. Microbe sand infection**, 2002.

BRACKETT, R. E. **Incidence, contributing factors and control of bacterial pathogens in produce.Postharvest Biology and Technology**, 1999.

CAMPOS, A. R, F.; SZEKUT, F. D.; KLEIN, M. R.; RIBEIRO, M. D. Application of treated sewage effluent applied to agriculture. **Inovagri International Meeting.**Fortaleza. Proceedings, 2015.

CARVALHO, P. H. Sweet bell pepper production in a protected environment with wastewater.**Green Magazine**, 2019.

CARVALHO, R. S .; SANTOS FILHO, J. S .; SANTANA, L. O. G.; GOMES, D. A .;MENDONÇA, L. C.; FACCIOLI, G. G. Influence of wastewater reuse on the microbiological quality of sunflower intended for animal feed. **Ambi-Agua**, 2013.

CASSIMIRO, C. A. L.; OLIVEIRA, F. S.; SILVA, E. A.; FEITOSA, S. S.; SIQUEIRA, E.C.; SILVA, M. G. Multiple water sheets via subsurface irrigation system in the cultivation of lettuce of the crespa group. **Revista brasileira de gestão ambiental** (Brazilian journal of environmental management, 2019.

COSTA, M. R. da S. Development of cabbage seedlings in different substrates and age**. Revista Verde de Agroecologia e Desenvolvimento Sustentável**, Pombal-PB, 2011.

DANTAS, I. L. A. Feasibility of using treated wastewater in irrigation of radish (Raphanus sativus L.) culture. **Revista Ambiente e Água**, 2014.

DOORENBOS, J.; PRUITT, W. O. **Crop water requirements**. 5 ed., Rome: FAO, 1977. 204 p. (FAO Studies, Irrigation and Drainage, 24).

DUARTE, A. **Reuse of treated wastewater for irrigation of chili crop**. Thesis (Doctorate in Agronomy). University of São Paulo, Piracicaba, 2006.

EMONGOR, V. E.; RAMOLEMANA, G. M. Treated sewage effluent (water) potential to be used for horticultural production in Botswana. **Physics and Chemistry of the Earth**, 2004.

FACCIOLI, G. G. **Determinação da evapotranspiriração de referência e da cultura da alface em condições de casa de vegetação, em Viçosa, MG.** Universidade Federal de Viçosa, 1998.

FAGGION, F.; OLIVEIRA, C. A. S.; CHRISTOFIDIS, D. Efficient use of water: a contribution to the sustainable development of agriculture and cattle ranching. **Applied Research and Agrotechnology**, 2009.

FEITOSA, S. O.; SILVA, S. L.; FEITOSA, H. O.; CARVALHO, C. M.; FEITOSA, E. O. Growth of cowpea bean irrigated with treated effluent and saline water under different concentrations. **Agropecuária Técnica**, 2015.

FELTRIM, A. L.; REGHIN, M. Y.; VAN DER VINNE, J. Cultivation of Pak Choi in different plant densities with and without nitrogen application.

Published by UEPG, Ponta Grossa, 2003.
FERREIRA, O. E.; BELTRÃO, N. E. M.; KONIG, A. Effects of wastewater and nitrogen application on growth and yield of herbaceous cotton. **Revista Brasileira de Oleaginosas e Fibrosas**, Campina Grande-PB, 2005.
FILGUEIRA, F. A. R. **Novo manual de olericultura: agrotecnologia moderna na produção e comercialização de hortaliças**. Universidade Federal de Viçosa, 2000.
FILGUEIRA, F. A. R. **Novo manual de olericultura: agrotecnologia moderna na produção e comercialização de hortaliças**. 3.ed. rev. e ampl. Viçosa, UFV, 2008.
FRANCO, B. D. G. M.; LANDGRAF, M. **Microbiologia de alimentos**. São Paulo: Atheneu, 1996.
FREITAS, C. A. S. Growth of sunflower crop irrigated with different types of water and nitrogen fertilization. **Revista Brasileira de Engenharia Agrícola e Ambiental**, 2012.
GOMES FILHO, R. R.; SANTOS, M. R. A.; CARVALHO, C. M.; FACCIOLI, G. G.; NUNES, T. P.; SANTOS, R. C.; VALNIR JÚNIOR, M.; LIMA, S. C. V.; MENDONÇA,M. C. S.; GEISENHOFF, L. O. Microbiological quality of lettuce irrigated with treated wastewater. **International Journal of Development Research,** 2020.

GONÇALVES, F. M.; FEITOSA, H. O.; CARVALHO, C. M.; GOMES FILHO, R. R.;VALNIR JUNIOR, M. Comparação de métodos da estimativa da evapotranspiriração de referência para o município de Sobral-CE. **Revista Brasileira de Agricultura Irrigada**, 2009.

HESPANHOL, I. Potential for water reuse in Brazil: agriculture, industry, municipalities, aquifer recharge. **Revista Brasileira de Recursos Hídricos** - RBRH, Porto Alegre, ed. Comemorativa, 2002.
HESPANHOL, I. Potential for water reuse in Brazil: agriculture, industry, municipality and aquifer recharge. In: MANCUSO, P. C. S. SANTOS, H. F. (Eds.). **Reuse of water.**Barueri, Manole, 2003.
JANG, T. I.; KIM, H. K.; SEONG, C. H.; LEE, E. J.; PARK, S. W. Assessing nutrient losses of reclaimed wastewater irrigation in paddy fields for sustainable agriculture. **Agricultural Water Management**, Amsterdam, 2012.
KERAITA, B.; JIMÉNEZ, B.; DRECHSEL, P. Extent and implications of agricultural reuse of untreated, partly treated and diluted wastewater in developing countries.**Perspectives in Agriculture, Veterinary Science, Nutrition and Natural Resources**, Wallingford, 2008.

LEFSRUD, Mark. Changes in kale (Brassica oleracea L. var. acephala) carotenoid and chlorophyll pigment concentrations during leaf ontogeny. **Scientia Horticulturae**, 2007.

LIMA, M. E.; CARVALHO, D. F.; SOUZA, A. P.; ROCHA, H. S.; GUERRA, J. G. M.Performance of eggplant cultivation in no-till submitted to different irrigation sheets. **Revista Brasileira de Engenharia Agrícola e Ambiental**, 2012.

LINO, S. R. L.; THIEL, A. A.; SILVA, R.; SOUZA, M. Reuse of water with a focus on family farming production. **Revista de Extensão do Instituto Federal Catarinense**, 2014.

LUCENA, C. Y. S.; SANTOS, D. J. R.; SILVA, P. L. S.; COSTA, E. D.; LUCENA, R. L. O Wastewater reuse as a means of coping with drought in the semi-arid northeastern Brazil. **Northeast Geosciences Journal**, 2018.

MAGNO, A. S. S. Influence of the use of treated wastewater on the agronomic characteristics of maxixe (Cucumis Anguria L.) **Scientific Initiation Scholarship Program UFS** 2018.

MANTOVANI, E. C.; BERNARDO, S.; PALARETTI, L. F. **Irrigation - Principles and Methods.** Viçosa: Editora UFV, 2nd Edition, 2007.

MARTÍNEZ, S.; SUAY, R.; MORENO, J.; SEGURA, M. L. **Reuse of tertiary municipal wastewater effluent for irrigation of Cucumis melo L,** 2012.

MASSARANDUBA, M. W. **Produção de cebola sob lâminas de irrigação e níveis de nitrogen na bacia do rio poxim**. Dissertation (Master in Hydric Resources) - Universidade Federal de Sergipe, 2020.

MEDEIROS, L. C. Morphometry of sunflowers irrigated with waste water and fertilized with different doses of nitrogen. **Brazilian Journal of Development,** 2020.

MOJID, M. A.; BISWAS, S. K.; WYSEURE, G. C. L. Interaction effects of irrigation by municipal wastewater and inorganic fertilizers on wheat cultivation in Bangladesh. **Field Crops Research**, Maryland, 2012.

MORENO, D. A.; CARVAJAL, M.; LOPEZ-BERENGUER, C.; GARCIA-VIGUERA, C. Chemical and biological characterization of nutraceutical compounds of broccoli**. Journal of Pharmaceutical and Biomedical Analysis**, 2006.

NOBRE, R. G. Sunflower growth irrigated with wastewater and organic fertilizer.**DAE Journal**, 2010.

NORETO, L. M.; MATTIELLO, V. D.; PARO, P.; KLEIN, J.; RICIERI, R. P.; SANTOS,
R. F.; FAGUNDES, R. S. Lettuce production subjected to different irrigation

fractions.Cultivating Knowledge, Cascavel, 2012.

NOVO, M. C.; PANTANO, A. P.; TRANI, P.; BLAT, S. Development and production of butter cabbage genotypes, state of São Paulo, Brazil, **Horticultura brasileira**, 2010.

OLIVEIRA, L. G. S. reuse of effluent in irrigation of sunflower (Helianthus annuus) Culture **Scholarship Program for Scientific Initiation - COPES/UFS**, 2013.

PAGANINI, W. S. **Reuse of water in agriculture.** In. MANCUSO P. C. S., SANTOS H. F. Reúso de água. Baureri, SP: Manole, 2003.

PALESE, A. M.; PASQUALE, V.; CELANO, G.; FIGLIUOLO, G.; MASI, S.; XILOYANNIS, C. Irrigation of olive groves in Southern Italy with treated municipal wastewa-ter: effects on microbiological quality of soil and fruits. **Agriculture, Ecosystems & Environment**, Amsterdam, 2009.

QADIR, M.; WICHELNS, D.; RASCHID-SALLY, L.; MCCORNICK, P. G.; DRECHSEL,P.; BAHRI, A.; MINHAS, P. S. The challenges of wastewater irrigation in developing countries. **Agricultural Water Management**, Amsterdam, 2010.

RAGAZZZI, M. F. **Estudo comparativo da qualidade parasitológica e toxicológica entre hortaliças cultivadas com água de reúso e hortaliças comercializadas em Ribeirão Preto - SP.** Ribeirão Preto, 2011.

RIBEIRO, M. S.; LIMA, L. A.; FARIA, F. H. DE. S.; REZENDE, F. C.; FARIA, L. do A. Effects of coffee wastewater on the vegetative growth of coffee trees in their first year**. Revista Engenharia Agrícola**, 2009.

REBOUÇAS, J. R. I. Crescimento do feijão-caupi irrigado com água residuária de esgoto doméstico tratado. **Revista Caatinga**, 2010.

REGO, J. de L.; OLIVEIRA, E. L. L.; CHAVES, A. F.; ARAUJO, A. P. B.; BEZERRA,
F. M. L.; SANTOS, A. B.; MOTA, S. **Revista Brasileira de Engenharia Agrícola e Ambiental**, 2005.

RODRIGUES, M. B, VILAS, M. A, SAMPAIO, S. C, REIS CF, GOMES SD. Effects of Fertirrigation with dairy and refrigerator wastewater on soil and refrigerator on soil and lettuce productivity. **Eng Ambient**, 2011.

RODRÍGUEZ-LIÉBANA, J. A.; ELGOUZIA, S.; MINGORANCEA, M. D.; CASTILLOA,
A.; PEÑA, A. Irrigation of a Mediterranean soil under field conditions with rbanwastewater: effect on pesticide behaviour. **Agriculture, Ecosystems and Environment**, Amsterdam, 2014.

SANTANA, J. S.; NASCIMENTO, C. H. S.; SILVA, C. M.; DAMASCENA, J. F. Response of lettuce cultivars under different irrigation and nitrogen doses. **Enciclopédia Biosfera**, v. 16, n. 29, p. 1332 - 1346.

SANTANA, F. S. Analise das influencias da utilização de água residuária tratada nas características agronômicas do coentro (Coriandrum sativum L.) **Programa de Bolsas de Iniciação Científica - COPES/UFS** 2019.

SANTOS, O. S. N.; PAZ, V. P. S.; GLOAGUEN, T. V.; TEIXEIRA, M. B.; FADIGAS, F. S.; COSTA, J. A. Crescimento e estado nutricional de helicônia irrigada com água residuária tratada em casa de vegetação. **Revista Brasileira de Engenharia Agrícola e Ambiental**, Campina Grande, 2012.

SANDRI, D.; MATSURA, E. E.; TESTEZLAF, R. Development of elisa lettuce in different irrigation systems with wastewater. **Brazilian Journal of Agricultural and Environmental Engineering**, 2007.

SILVA, N.; **Manual de Métodos de Análise Microbiológica de Alimentos e Água.** 4ª edição. São Paulo. Editora Varela, 2010.

SILVA, M. B. R.; FERNADES, P. D.; DANTAS NETO, J.; NERY, A. R.; RODRIGUES,L. N.; VIEGAS, R. A. Growth production of jatropha irrigated with wastewater under water stress conditions. **Revista Brasileira de Engenharia Agrícola e Ambiental,** 2011.

SILVA, V. F.; NASCIMENTO, E. C. S.; ANDRADE, L.O.; BARACUHY, J. G. V; LIMA,V.L.A. Effect of bovine substrate on germination of biquinho pepper (Capsicum chinense) irrigated with wastewater. **Journal Environmental Monographs**, 2014.

SILVA, L. S. Response to microbiological water treated domestic irrigation characteristics of beans grown in a protected environment. **International journal of development research**, 2020.

SILVA, E. L. Use of treated domestic sewage in the development of cowpea crop in a protected environment. **Dissertation-prorh**, 2019

SOBRAL, L. F.; VIEGAS, P. R. A.; SIQUEIRA, O. J. W.; ANJOS, J. L.; BARRETTO, M.C. de V.; GOMES, J. B. V. **Recomendação para o uso de corretivos e fertilizantes no Estado de Sergipe.** Aracaju: Embrapa Tabuleiros Costeiros, 2007. 251 p.

SOUZA, F. M. R. Viabilidade no uso de água residuária tratada na irrigação da cultura da beterraba (beta vulgaris l.) e da carro carrot (daucus carota l.). **Programa de Bolsas de Iniciação Científica - COPES/UFS** 2014.

SOUZA, F. M. R. Analise das características microbiológicas do Feijão caupi cultivar brs novaera irrigado com água residuária tratada, **Programa de Bolsas de Iniciação Científica- COPES/UFS** 2015.

SOUZA, F. M. R. Influence of the use of treated wastewater on the agronomic characteristics of okra (Abelmoschus esculentus L), Scientific Initiation Scholarship Program - COPES/UFS 2017.
SOUZA, F. M. R Influence of the use of treated wastewater on the agronomic characteristics of bell bell pepper. (Capsicum annuum L.), Scientific Initiation Scholarship Program - COPES/UFS, 2016.
TRANI, P. E. Leafy cabbage from planting to post-harvest. Boletim técnico. Agronomic Institute of campinas, 2015.
XAVIER, J. F. Cultivo da mamoneira sob diferentes tipos de águas residuárias e de abastecimento e níveis de água no solo. Revista Caatinga, Mossoró, 2014.
WANDERLEY, T. F. Avaliação dos Efeitos do Reúso de Águas de esgotos sobre a Produtividade e a Qualidade Microbiológica de Cultivares de Batata-doce visando à Produção de Biomassa. 2005. 130 f. Dissertation (Master in Environmental Sciences) - Federal University of Tocantins. Tocantins: 2005.
WHO, Wastewater if in Agriculture, in: Guidelines for the Safe Use of Wastewater, Excreta and Greywater, vol.2, World Health Organization, Geneva, Switzerland, 2006.
YAMAGUCHI, M. U.; ZANQUETA, E. B.; MOARAIS, J. F.; FRAUSTO, H. S. E. G.;SILVERIO, K. I. Microbiological quality of food and work environments: research of Salmonella and Listeria. Journal in agribusiness and environment, 2013.

ANNEXES

INSTITUTO TECNOLÓGICO E DE PESQUISAS DO ESTADO DE SERGIPE

Rua Campo do Brito, Nº371, Treze de Julho, CEP 49.020-380
Aracaju - SE - Brasil

Fone (79) 3179-8081/8087 Fax (79) 3179-8087/8090
CNPJ 07.258.529/0001-59

Relatório de Ensaios ITPS Nº 3199/19-1

Revisão 00

Cliente	ANGELIS CARVALHO MENEZES	**Telefone**	79 9 9857-6790
Endereço	Rua Tenente Antônio Fontes Pitanga, bloco felicita, apto. 401, 256, CEP 49032-360	**Contato(s)**	ANGELIS CARVALHO MENEZES
e-mail	angelis.menezes@gmail.com	**Fax**	
Amostra(s)	Solo	**Recepção**	29/07/19

Laboratório de ensaios acreditado pela norma ABNT NBR ISO/IEC 17025:2005

O escopo da acreditação pode ser visto em:
http://www.inmetro.gov.br/laboratorios/rble/docs/CRL0424.pdf

Amostra	SOLOS CAMPUS RURAL DA UFS		**Código**	3199/19-01	**Coleta em** –
Ensaio	**Resultado**	**Unidade**	**LQ**	**Método**	**Data do Ensaio**
pH em Água (RBLE)	**5,93**	--	--	H_2O	07/08/19 00:00
Cálcio + Magnésio (RBLE)	**1,13**	cmolc/dm3	0,38	MAQS-Embrapa 2009, KCl	07/08/19
Cálcio (RBLE)	**0,77**	cmolc/dm3	0,22	MAQS-Embrapa 2009, KCl	07/08/19
Alumínio (RBLE)	**<0,08**	cmolc/dm3	0,08	MAQS-Embrapa 2009, KCl	07/08/19
Sódio (RBLE)	**3,40**	mg/dm3	2,20	MAQS-Embrapa 2009, Mehlich-1	07/08/19
Potássio (RBLE)	**26,7**	mg/dm3	1,40	MAQS-Embrapa 2009, Mehlich-1	07/08/19
Fósforo (RBLE)	**6,00**	mg/dm3	1,39	MAQS-Embrapa 2009, Mehlich-1	07/08/19

Coleta efetuada pelo cliente.
A descrição do material ensaiado é de inteira responsabilidade do cliente.

Aracaju, 08 de agosto de 2019.

Rivaldo Cordeiro Santos
Eng. Agrônomo
CREA-SE 1.308
Química Agrícola

Documento verificado e aprovado por meios eletrônicos
A verificação da autenticidade deste documento pode ser feita baixando o documento original em www.itps.se.gov.br na aba Serviços clicando em Resultados de Análises usando o código LTDGZ CD9 529.

A Custódia das amostras é de 15 dias após emissão do relatório de ensaios, exceto para solos que é 90 dias e água que é 2 dias. Não se aplica a amostras perecíveis. Os resultados têm significado restrito e aplicam-se somente às amostras ensaiadas. Este relatório somente poderá ser reproduzido em sua totalidade. O ITPS se isenta de qualquer responsabilidade pela reprodução parcial do mesmo.

RF-LBW-004, Rev. 00 **Página: 1/1**

INSTITUTO TECNOLÓGICO E DE PESQUISAS DO ESTADO DE SERGIPE

Rua Campo do Brito, Nº371, Treze de Julho, CEP 49.020-380 Aracaju - SE - Brasil

Fone (79) 3179-8081/8087 Fax (79) 3179-8087/8090 CNPJ 07.258.529/0001-59

Relatório de Ensaios ITPS Nº 3199/19-2

Revisão 00

Cliente	ANGELIS CARVALHO MENEZES	**Telefone**	79 9 9857-6790
Endereço	Rua Tenente Antônio Fontes Pitanga, bloco felicita, apto. 401, 256, CEP 49032-360	**Contato(s)**	ANGELIS CARVALHO MENEZES
e-mail	angelis.menezes@gmail.com	**Fax**	
Amostra(s)	Solo	**Recepção**	29/07/19

Os Resultados relatados abaixo não fazem parte do escopo da acreditação deste Laboratório

Amostra	SOLOS CAMPUS RURAL DA UFS		**Código**	3199/19-01	**Coleta em** –
Ensaio	**Resultado**	**Unidade**	**LQ**	**Método**	**Data do Ensaio**
Matéria Orgânica	**9,17**	g/dm3	--	WB (colorimétrico)	07/08/19
Magnésio	**0,36**	cmolc/dm3	--	MAQS-Embrapa 2009, KCl	08/08/19
Sódio	**0,015**	cmolc/dm3	--	MAQS-Embrapa 2009, Mehlich-1	08/08/19
Potássio	**0,07**	cmolc/dm3	--	MAQS-Embrapa 2009, Mehlich-1	08/08/19
Hidrogênio + Alumínio	**0,857**	cmolc/dm3	--	SMP	07/08/19
pH em SMP	**7,2**	--	--	MAQS-Embrapa	07/08/19
SB-Soma de Bases Trocáveis	**1,22**	cmolc/dm3	--	--	08/08/19
CTC	**2,08**	cmolc/dm3	--	--	08/08/19
PST	**0,72**	%	--	--	08/08/19
V - Índice de Saturação de Bases	**58,7**	%	--	--	08/08/19

Legenda

MAQS-Embrapa: Manual de Análises Químicas de Solos, Plantas e Fertilizantes, Embrapa 2009. Análise realizada em amostra de terra fina seca em estufa (t.f.s.e.) a 40ºC. Conversão de Unidades: cmolc/dm3=meq/100g; g/dm3=% X 10; % = dag Kg-1.

LQ: Limite de Quantificação do Método.

Informações de Coleta

Coleta efetuada pelo cliente.
A descrição do material ensaiado é de inteira responsabilidade do cliente.

Aracaju, 08 de agosto de 2019.

Rivaldo Cordeiro Santos
Eng. Agrônomo
CREA-SE 1.308
Química Agrícola

Documento verificado e aprovado por meios eletrônicos
A verificação da autenticidade deste documento pode ser feita baixando o documento original em www.itps.se.gov.br na aba Serviços clicando em Resultados de Análises usando o código LTDGZ CD9 529.

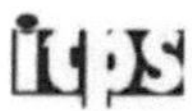

INSTITUTO TECNOLÓGICO E DE PESQUISAS DO ESTADO DE SERGIPE

Rua Campo do Brito, Nº371, Treze de Julho, CEP 49.020-380
Aracaju - SE - Brasil

Fone (79) 3179-8081/8087 Fax (79) 3179-8087/8090
CNPJ 07.258.529/0001-59

Relatório de Ensaios ITPS Nº 0306/20

Revisão 00

Cliente	MARCOS HENRIQUE RAMOS MENEZES	**Telefone**	79 9 9858-1857 FRAN
Endereço	CLAUDIO BATISTA AP 204 BL2, 295	**Contato(s)**	FUNC ITPS
e-mail		**Fax**	
Amostra(s)	Águas MB	**Recepção**	22/01/20

Amostra	T1			**Código**	0306/20-01	**Coleta em**	22/01/20
Ensaio	**Resultado**	**Unidade**	**Padrão (L1)**	**LQ**	**Método**		**Data do Ensaio**
Salmonellas	**Ausência**	em 25g	Ausência	--	AOACC 967.26		22/01/20
Coliformes a 45°C	**<3,0**	NMP/g	10^2	--	SMEWW9221B		22/01/20

Conclusão dos Ensaios (Parecer Técnico*): De acordo com os parâmetros analisados para o atendimento de "RDC nº 12/2001 da ANVISA", os resultados reportados neste relatório para esta amostra **atendem** aos limites estabelecidos.

Legenda

(L1): RDC nº 12/2001 da ANVISA
NMP: Número Mais Provável.
SMEWW: Standard Methods for the Examination of Dairy Products, APHA, 17ª. ed.,Washington, 2004
Resultado: Resultados fora de faixas aparecem sublinhados.
LQ: Limite de Quantificação do Método.
Parecer Técnico*: Os pareceres, interpretações e opiniões expressos não fazem parte do escopo do sistema de qualidade deste laboratório com base na norma NBR ISO/IEC 17025.

Informações de Coleta

Coleta efetuada pelo cliente.
A descrição do material ensaiado é de inteira responsabilidade do cliente.

MESTRADO.

Aracaju, 03 de fevereiro de 2020.

Douglas Bonfim Lima
Biólogo

Documento verificado e aprovado por meios eletrônicos
A verificação da autenticidade deste documento pode ser feita baixando o documento original em www.itps.se.gov.br na aba Serviços clicando em Resultados de Análises usando o código LUCNZ DBZ 149.

INSTITUTO TECNOLÓGICO E DE PESQUISAS DO ESTADO DE SERGIPE

Rua Campo do Brito, Nº371, Treze de Julho, CEP 49.020-380 Aracaju - SE - Brasil
Fone (79) 3179-8081/8087 Fax (79) 3179-8087/8090
CNPJ 07.258.529/0001-59

Relatório de Ensaios ITPS Nº 0306/20 **Revisão** 00

Cliente	MARCOS HENRIQUE RAMOS MENEZES	Telefone	79 9 9858-1857 FRAN
Endereço	CLAUDIO BATISTA AP 204 BL2, 295	Contato(s)	FUNC ITPS
e-mail		Fax	
Amostra(s)	Águas MB	Recepção	22/01/20

Amostra T2				Código	0306/20-02	Coleta em 22/01/20
Ensaio	**Resultado**	**Unidade**	**Padrão (L1)**	**LQ**	**Método**	**Data do Ensaio**
Salmonellas	**Ausência**	em 25g	Ausência	--	AOACC 967.26	22/01/20
Coliformes a 45°C	**<3,0**	NMP/g	10^2	--	SMEDP9221B	22/01/20

Conclusão dos Ensaios (Parecer Técnico*): De acordo com os parâmetros analisados para o atendimento de "RDC nº 12/2001 da ANVISA", os resultados reportados neste relatório para esta amostra **atendem** aos limites estabelecidos.

Legenda
(L1): RDC nº 12/2001 da ANVISA
NMP: Número Mais Provável.
SMEWW: Standard Methods for the Examination of Dairy Products, APHA, 17ª. ed.,Washington, 2004
Resultado: Resultados fora de faixas aparecem sublinhados.
LQ: Limite de Quantificação do Método.
Parecer Técnico*: Os pareceres, interpretações e opiniões expressos não fazem parte do escopo do sistema de qualidade deste laboratório com base na norma NBR ISO/IEC 17025.

Informações de Coleta

Coleta efetuada pelo cliente.
A descrição do material ensaiado é de inteira responsabilidade do cliente.

MESTRADO.

Aracaju, 03 de fevereiro de 2020.

Douglas Bonfim Lima
Biólogo

Documento verificado e aprovado por meios eletrônicos
A verificação da autenticidade deste documento pode ser feita baixando o documento original em www.itps.se.gov.br na aba Serviços clicando em Resultados de Análises usando o código LUCNZ DBZ 149.

A Custódia das amostras é de 15 dias após emissão do relatório de ensaios, exceto para solos que é 90 dias e água que é 2 dias. Não se aplica a amostras perecíveis. Os resultados têm significado restrito e aplicam-se somente às amostras ensaiadas. Este relatório somente poderá ser reproduzido em sua totalidade. O ITPS se isenta de qualquer responsabilidade pela reprodução parcial do mesmo.

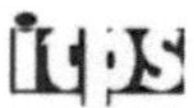

INSTITUTO TECNOLÓGICO E DE PESQUISAS DO ESTADO DE SERGIPE

Rua Campo do Brito, Nº371, Treze de Julho, CEP 49.020-380
Aracaju - SE - Brasil

Fone (79) 3179-8081/8087 Fax (79) 3179-8087/8090
CNPJ 07.258.529/0001-59

Relatório de Ensaios ITPS Nº 0306/20

Revisão 00

Cliente	MARCOS HENRIQUE RAMOS MENEZES	**Telefone**	79 9 9858-1857 FRAN
Endereço	CLAUDIO BATISTA AP 204 BL2, 295	**Contato(s)**	FUNC ITPS
e-mail		**Fax**	
Amostra(s)	Águas MB	**Recepção**	22/01/20

Amostra	T3			**Código**	0306/20-03	**Coleta em** 22/01/20
Ensaio	**Resultado**	**Unidade**	**Padrão (L1)**	**LQ**	**Método**	**Data do Ensaio**
Salmonellas	**Ausência**	em 25g	Ausência	--	AOACC 967.26	22/01/20
Coliformes a 45°C	**<3,0**	NMP/g	10^2	--	SMEDP9221B	22/01/20

<u>**Conclusão dos Ensaios (Parecer Técnico*):**</u> De acordo com os parâmetros analisados para o atendimento de "RDC nº 12/2001 da ANVISA", os resultados reportados neste relatório para esta amostra **atendem** aos limites estabelecidos.

Legenda
(L1): RDC nº 12/2001 da ANVISA
NMP: Número Mais Provável.
SMEWW: Standard Methods for the Examination of Dairy Products, APHA, 17ª. ed.,Washington, 2004
Resultado: Resultados fora de faixas aparecem sublinhados.
LQ: Limite de Quantificação do Método.
Parecer Técnico*: Os pareceres, interpretações e opiniões expressos não fazem parte do escopo do sistema de qualidade deste laboratório com base na norma NBR ISO/IEC 17025.

Informações de Coleta

Coleta efetuada pelo cliente.
A descrição do material ensaiado é de inteira responsabilidade do cliente.

MESTRADO.

Aracaju, 03 de fevereiro de 2020.

Douglas Bonfim Lima
Biólogo

Documento verificado e aprovado por meios eletrônicos
A verificação da autenticidade deste documento pode ser feita baixando o documento original em www.itps.se.gov.br na aba Serviços clicando em Resultados de Análises usando o código LUCNZ DBZ 149.

Printed by Books on Demand GmbH, Norderstedt / Germany